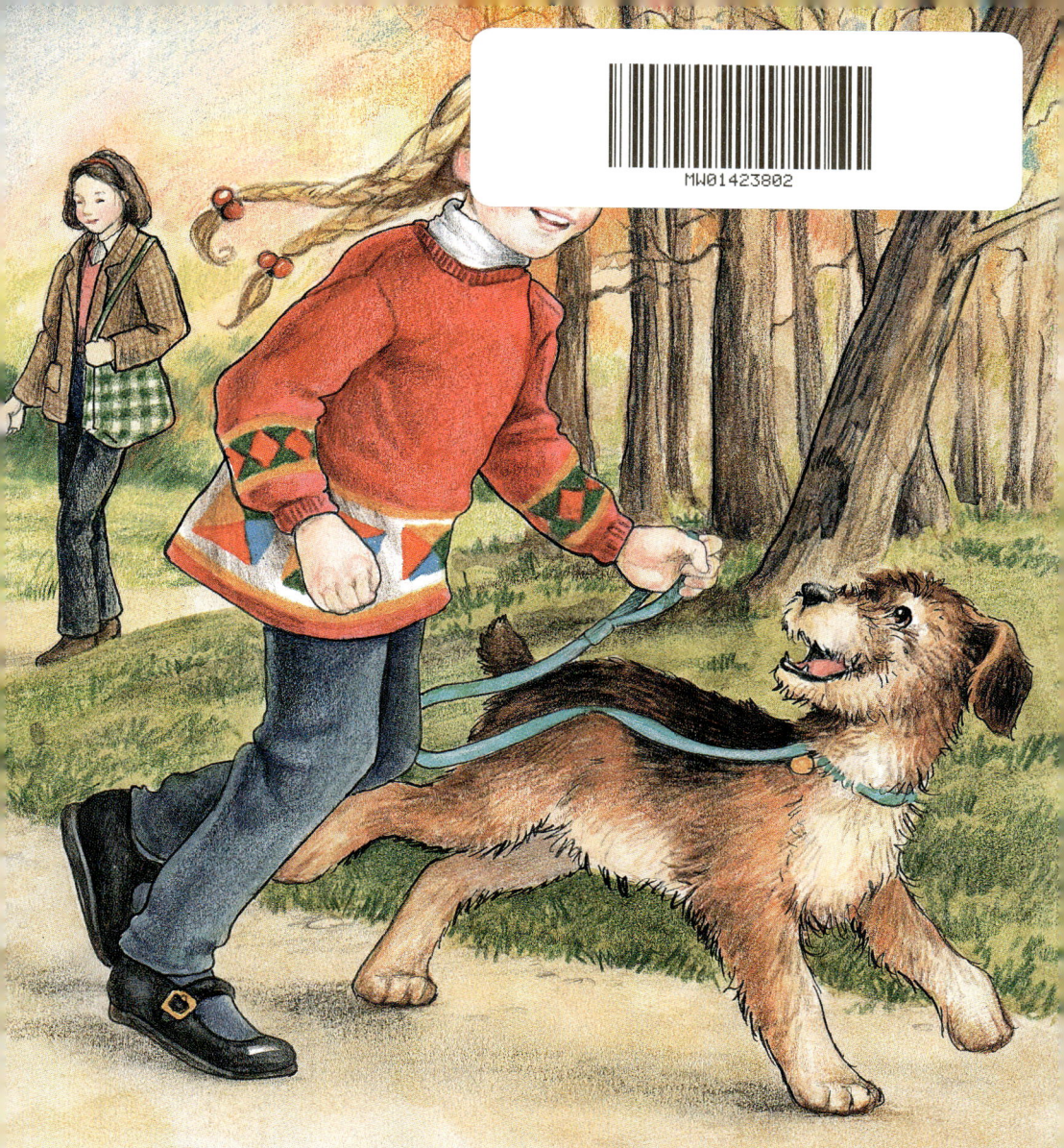

"I love going to the park!" said Jill.

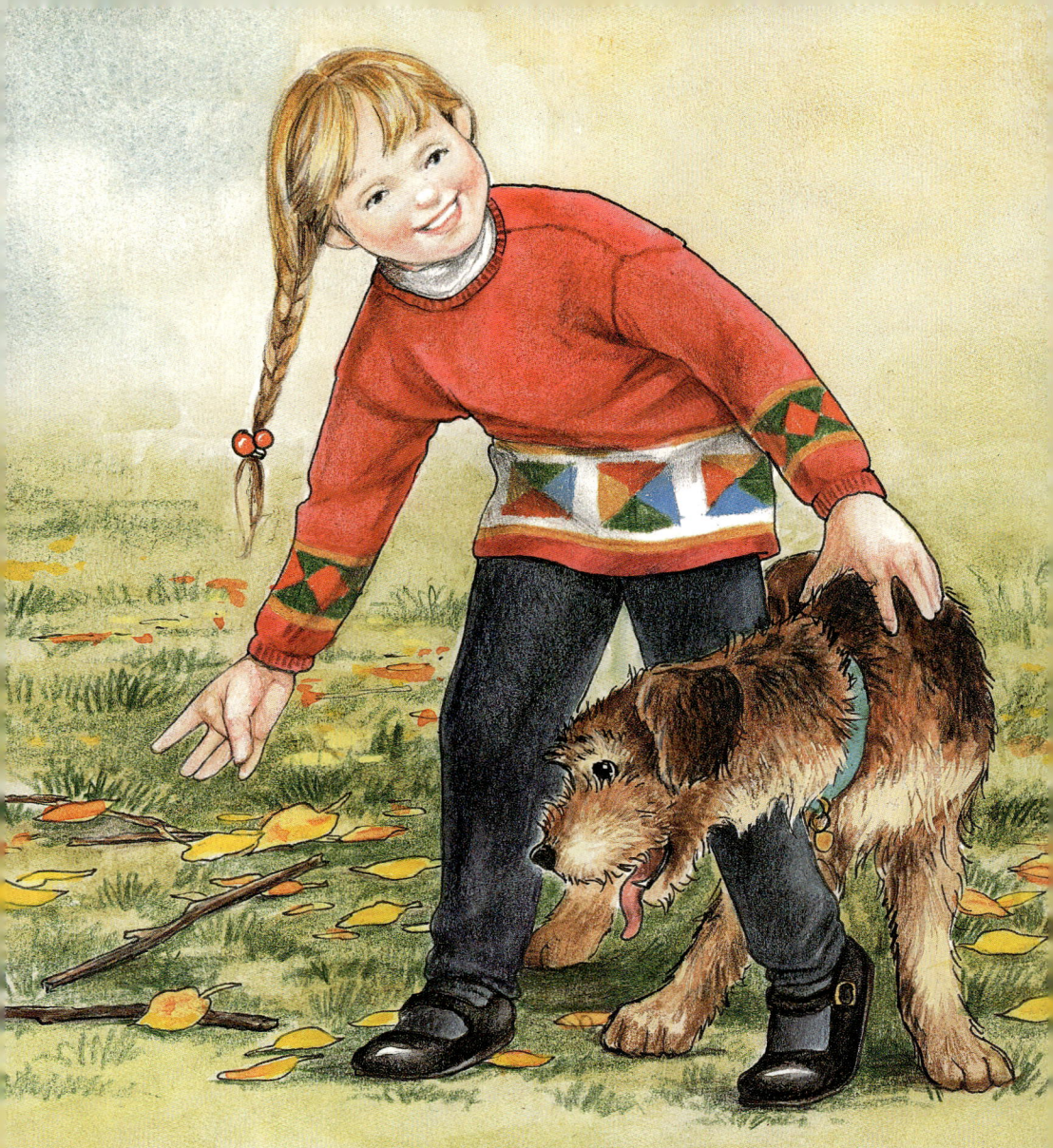

Jill picked up sticks for Ned.

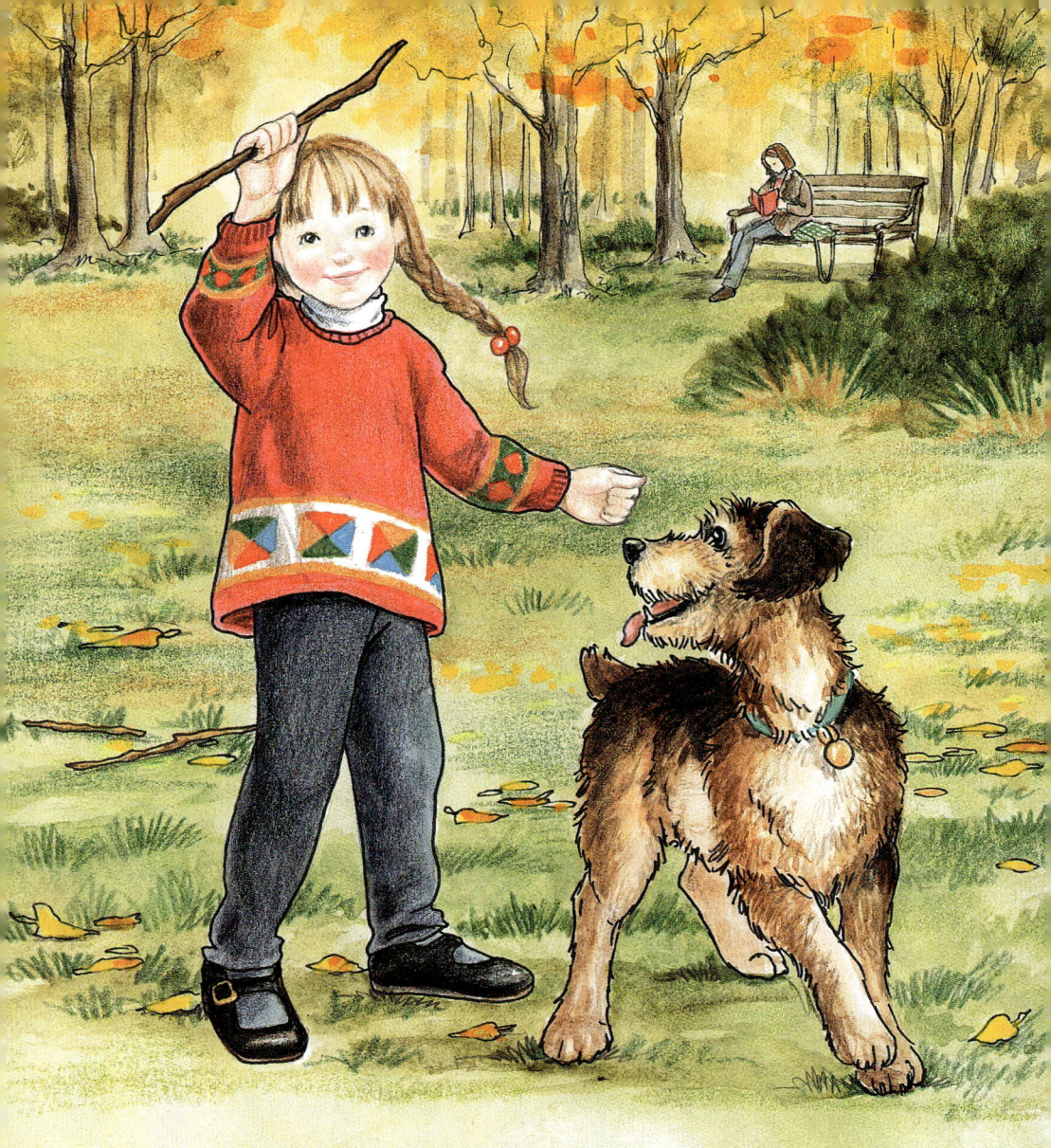

Jill tossed a stick to him.

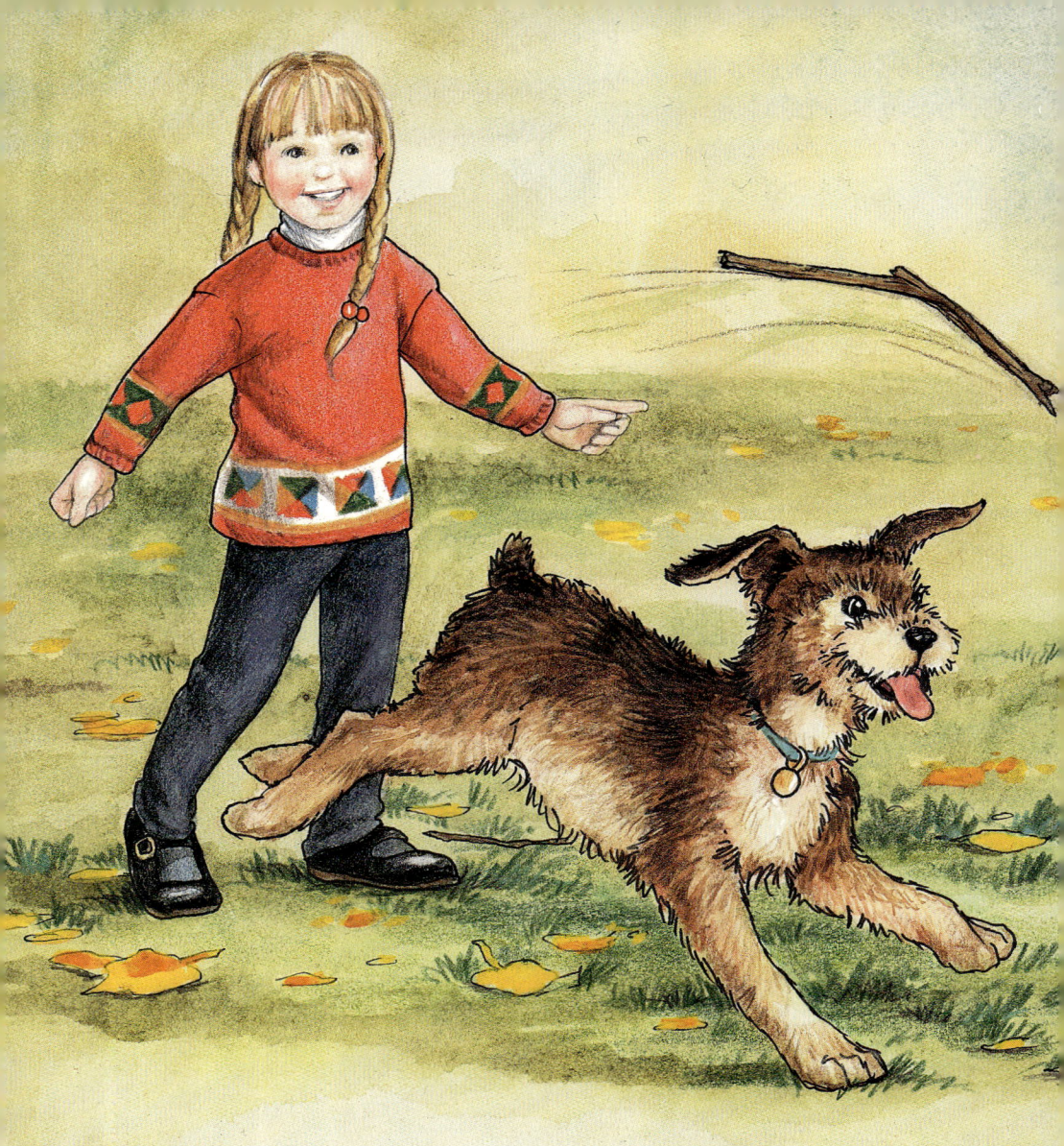

Ned started to run fast.

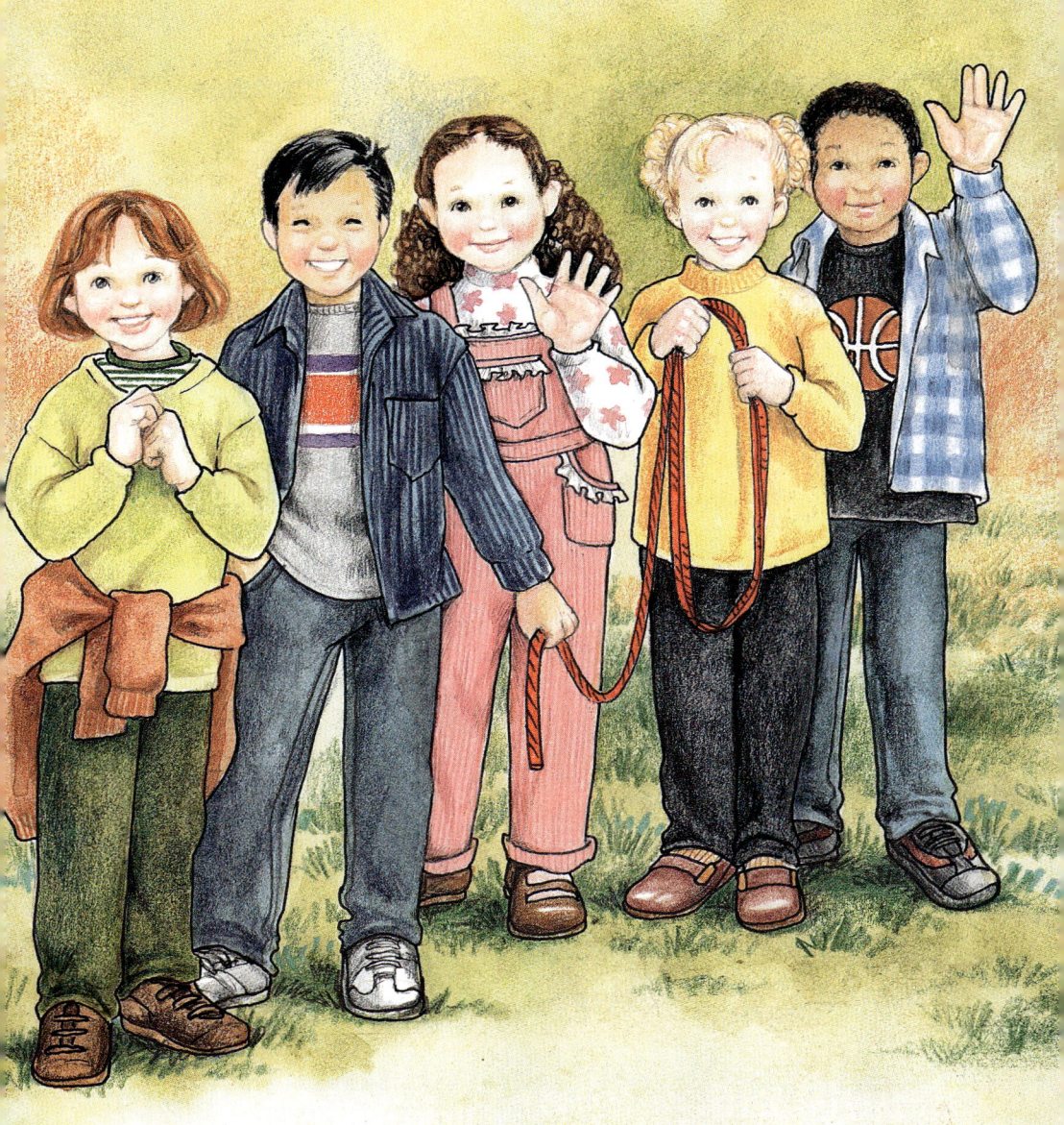

Some pals yelled, "Come with us!"

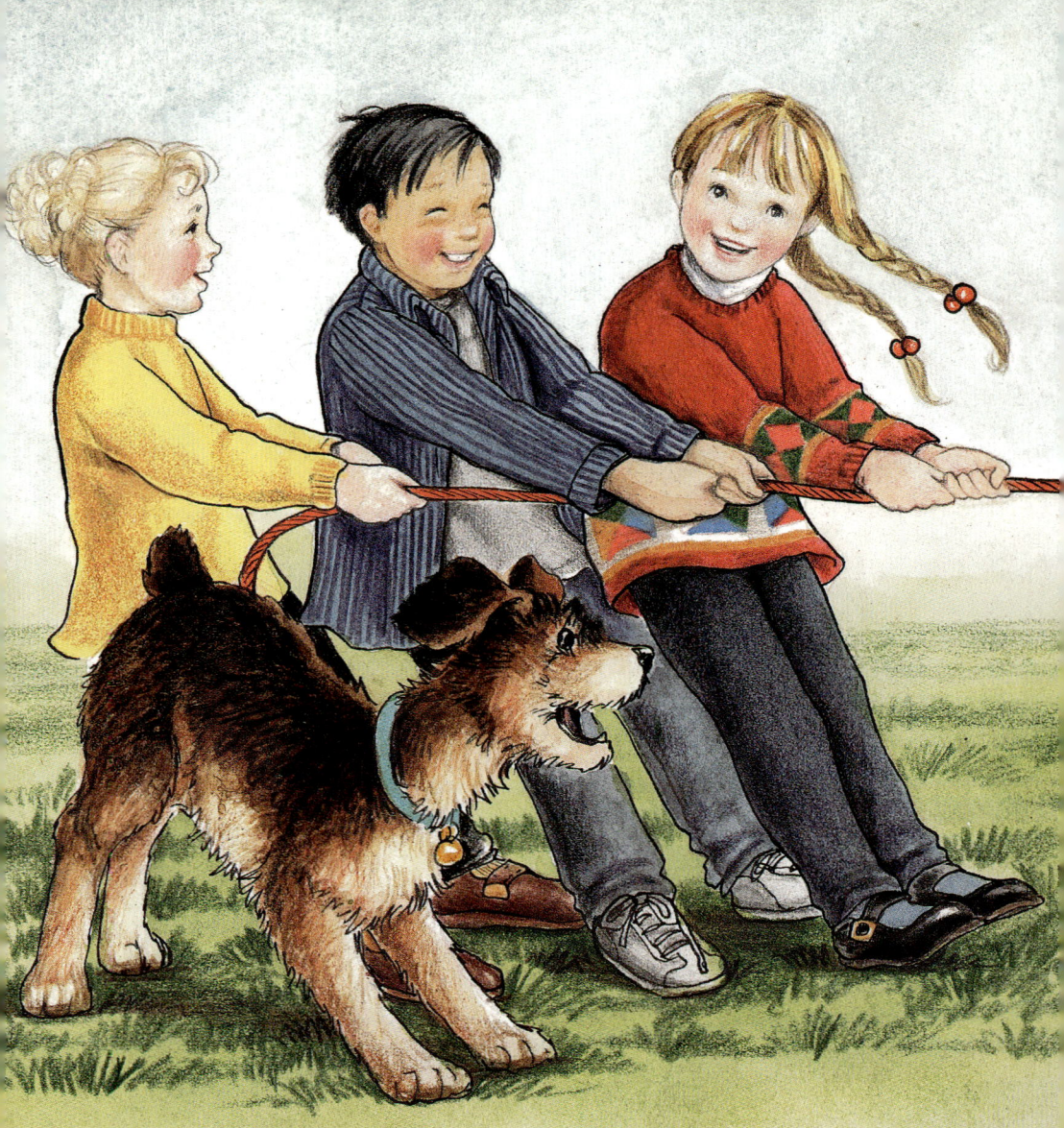
"We are backing up!" Jill yelled.

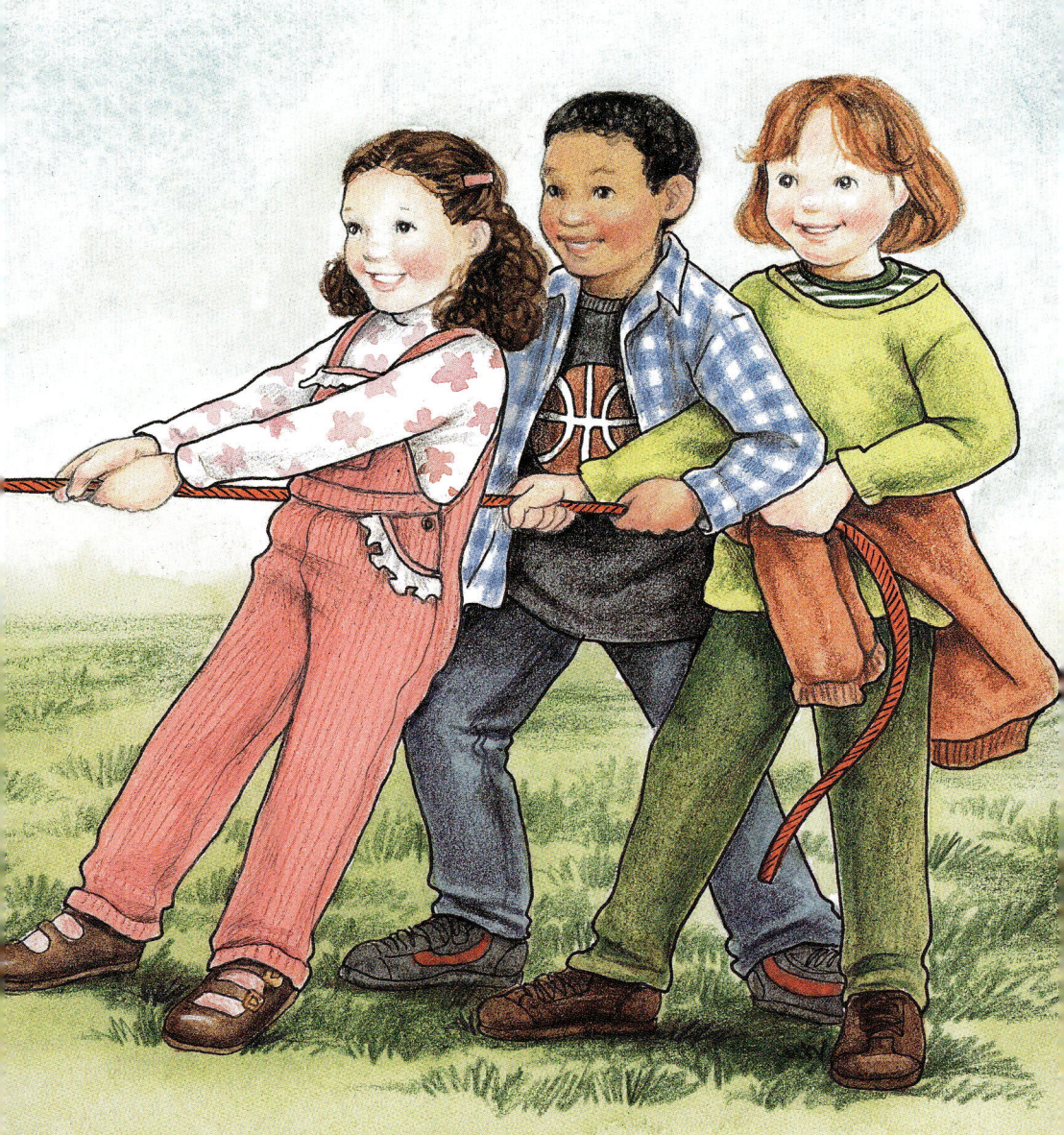

Then they backed up!

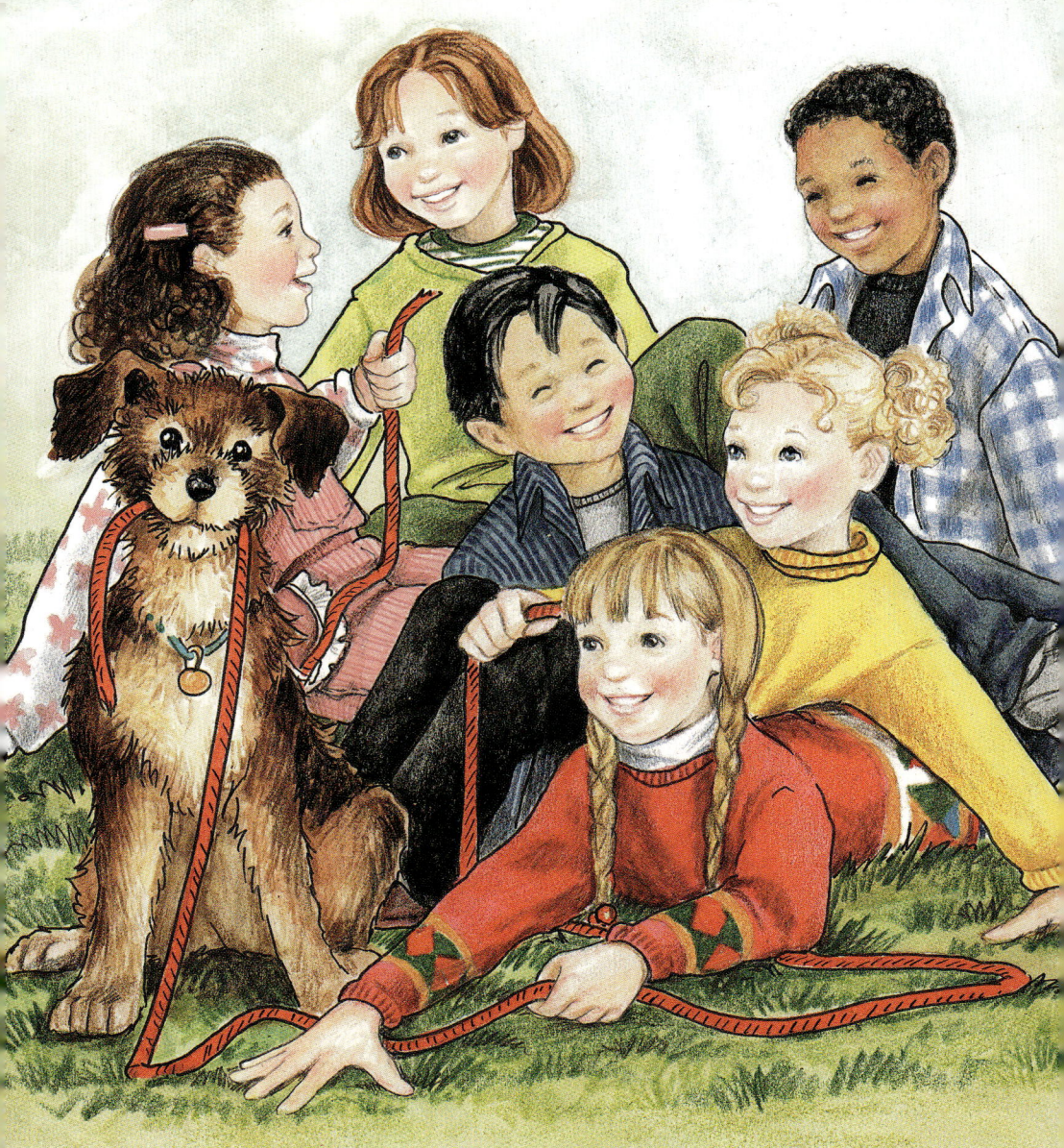

Who fell?